I0475071

GO SOLAR AND SAVE DOLLARS

1ST EDITION

Introductory guide to the world of solar energy and its practical applications.

Siemens, Michael. Delfin

medical or professional advice. The content of this book has been derived from various sources. Please consult a licensed professional before attempting any techniques outlined in this book.

By reading this document, the reader agrees that under no circumstances is the author responsible for any losses, direct or indirect, which are incurred as a result of the use of information contained within this document, including, but not limited to, —errors, omissions, or inaccuracies.

Stupid and sensual laziness... Stop seducing me

Contents

Introduction

The sun is the primary source of energy on Earth and sunlight can be converted directly into electricity using solar panels. Electricity has become indispensable in life. It powers the machines that most us use daily.

So, what are solar panels? What if you can create your own?

In this book, we will show you a straightforward method of building your own functional solar panel.

THE COMPONENTS

A solar panel is usually manufactured from six (6) components namely the PV (photovoltaic) cell or solar cell which generates the electricity, the glass which covers and protects the solar cells, the frame which provides rigidity, the backsheet where the solar cells are laid, the junction box where the wirings are enclosed and connected, and the encapsulant which serves as adhesives.

Since most people does not have access to e□uipment in manufacturing solar panels, it is important to note and understand those six components in order for anyone to be able to plan the materials needed to create a do-it-yourself or home-made solar panel.

The materials needed on how to make a solar panel must be available for purchase locally or online and should not exceed the cost of a brand new solar panel or does not take a long time to build.

1.) PV Cell

The first thing to consider when building your own solar panel is the solar cell.

Photovoltaic (PV) cell or solar cell converts visible light into electricity. One (1) solar cell however is not enough to produce a usable amount of electricity much like the microbot in Baymax (Hero 6) which only becomes useful when combined as a group. This basic unit generates a DC (direct current) voltage of 0.5 to 1 volt and although this is reasonable, the voltage is still too small for most applications. To produce a useful DC voltage, the solar cells are connected in series and then encapsulated in modules making the solar panel. If one cell generates 0.5 volt and is connected to another cell in series, those two cells should then be able to produce 1 volt and they can then be called a module. A typical module usually consists of 28 to 36 cells in series. A 28-cell module should be able to produce roughly 14 volts (28 x 0.5 = 14VDC) which is enough to charge a 12V battery or power 12V devices.

Connecting two or more solar cells require that you have a basic understanding of series and parallel connection which is similar to connecting batteries to make up a battery storage system.

There are two most common solar cells that can be bought in the market; a monocrystalline cell and a polycrystalline cell. These two can have the same size, 156mm x 156mm, but the main difference would be efficiency. It is important to purchase additional cells to serve as backup in case you fail on some of the cells i.e. bad solder, broken cell, scratched, etc.

Monocrystalline solar cells are usually black and octagonal in shape. This type of solar cell is made of the highest and purest grade silicon which makes them expensive. But they are the most efficient of all types of solar cells and are almost always the choice of solar contractors when space is an important factor to consider in achieving the power they want to attain based on their solar system design.

Polycrystalline PV Cells are characterized by their bluish color and rectangular shape. These cells are manufactured in a much simpler process which lowers the purity of the silicon content and also lowers the efficiency of the end product.

Generally, monocrystalline cells are more efficient than polycrystalline cells but this does not mean that monocrystalline cells perform and outputs more power than polycrystalline cells. Solar cell efficiency has something to do with the size of the cells and every solar panel or cells have an efficiency rating based on standard tests when they were manufactured. This rating is usually in percentage and the common values range from 15% to 20%.

2.) Glass

The glass protects the PV cells while allowing optimal sunlight to pass through. These are usually made of anti-reflective materials. Tempered glass is the choice of material nowadays even for unknown and new manufacturers although there are still those who utilize flat plate glass on their solar panels. Tempered glass are created by chemical or thermal means and is many times stronger than plate glass making it more expensive to produce but the price of manufacturing them today is reasonable and cost-effective. Flat plate glass creates sharp and long shards when broken as opposed to tempered glass which shatters safely in small pieces upon impact, that is why they also call it safety glass. It should be noted here that most amorphous solar panels uses flat plate glass because of the way the panel is constructed.

Tempered glass is what manufacturers use in mass producing their solar panels. In our DIY project, we suggest to use Plexiglas also called acrylic glass which is safer than the regular normal glass from your local hardware store. It is a bit expensive than regular glass but is weather resistant and does not break easily. The Plexiglas can also be screwed or glued easily to the frame.

3.) Frame

A frame is usually made of anodized aluminum which provides structure and rigidity to the solar module.

These aluminum frames are also designed to be compatible with most solar mounting systems and grounding equipment for easy and safe installation on a roof or on the ground.

The frame in a factory-built solar panel is usually the aluminum part where all four sides of the solar panel sheet are inserted. Think of it as a skeletal rectangular frame. The solar panel sheet by the way is composed of the other 4 components and are layered and laminated in the following order from top to bottom; the tempered glass, top encapsulant, the solar cells, bottom encapsulant, then the backsheet. In our DIY solar panel, we will be using a wooden frame and the end-result would be something analogous to a picture frame where the picture is the solar cells glued to a non-conductive board, the glass for the Plexiglas top cover, and the wooden part as the frame and backsheet.

4.) Backsheet

The backsheet is the layer of plastic film on the back surface of the module. This is the only layer protecting the module from unsafe DC voltage. The main function of the backsheet is to insulate and protect the handler from shock and provide the safest, efficient, and dependable electrical conductivity possible.

The backsheet will be a wooden plywood where the frame will be screwed on top and on the sides. It should be noted here that a perforated hardboard (Pegboard) will be used to place and align the PV Cells and this Pegboard will sit on top of the wooden backsheet and fitted inside the wooden frame.

5.) Junction Box

The junction box is where the terminal wires and bypass diodes are located and concealed. The terminal wires are basically the positive and negative wires based on the series connections of the PV Cells and can be connected to another solar panel, a charge controller, a battery system, or to an inverter, depending on the system design. The bypass diode is a protective mechanism that prevent power from getting back to the solar panel when it is not producing electricity as in the case when it is night time.

There are junction boxes designed for factory-built solar panels that are now available to purchase online especially from China. If you are not pressed for time, you can order online and wait for the delivery otherwise you can just purchase a regular electrical junction box from your local hardware store. The purpose of the junction box is to protect the terminals (positive and negative terminals) from water, dust, and other elements. This is also where the two wires (red for positive and black for negative) will be coming from. The other end of these two wires can also be protected by using a PV accessory called MC4 which can also be purchased online together with the PV junction box.

6.) Encapsulant

Encapsulant sheets prevent water and dirt from infiltrating the solar modules and serve as shock-absorbers that protect the PV cells. They have this adhesive bonding capability to the glass, the PV cells, and the backsheet similar to a glue but stronger. Encapsulants are usually made of Ethylene-Vinyl

Acetate or EVA and are applied using lamination machines and processes. Solar panel manufacturers use a vacuum and a large oven to properly seal and cure the EVA sheet onto the solar panels. Most of us do not have the capability to do this but many still have tried and failed while others had varying levels of success.

Encapsulants are thin plastic sheets that are usually laminated on the top and bottom parts of the solar cell sheet. The bottom encapsulant is the layer on top of the backsheet where the solar cells are actually placed and supported. In our project, we will instead use a latex acrylic paint. This paint will not be applied to the pv cells because when attempted, will not result in an e□ual distribution or application of the liquid to the surface of the cells which can degrade performance. The paint will be applied to the wooden frame, wooden backsheet, and to the Pegboard. This Latex Acrylic paint should be able to protect the wooden parts from UV rays and can better resist blistering and cracking overtime. This paint, although water-soluble, can dry □uickly and becomes water-resistant.

Chapter One

Use Solar Panels to Save Money and the Environment

The notion of unlimited and free energy has naturally attracted the interest of the environmentally conscious and frugal to use solar panels to save money and the environment. You can begin and enter this world by building panels yourself. You will need to use a guide book and a kit that provides easy to follow instructions and includes videos as an aid. You will have to do your research to find what works for you. When used properly these kits are a great learning opportunity and help you power simple devices. More kits are coming on the market as there is enormous demand for trying this route to energy production. For those who do not want to go all the way yet, this is a good place to start. Solar energy production techniques are being refined and improved on, so it may be wise not to make major investments at this time. In fact as revealed in the British media in February 2010, scientists have discovered a plastic that can store and release energy and may make batteries obsolete.

Solar panels are a collection of solar cells wired together. They produce an electrical current when struck by sunlight. Although, solar cells generate minimal electricity individually, grouped together they generate substantially more. You can save money by building them yourselves rather than purchase them prefabricated. You can easily purchase solar cells you will wire together to make a panel online. You can purchase good quality whole cells, or lesser ꋚuality cells that are cheaper but are blemished in some way. Pre-tabbed cells facilitate wiring. You can save up to half to three quarters of the cost, if you make the panel yourself compared to a premade panel. If you purchase the wood and pretabbed cells a panel can cost a little over 100 USD for a small system for a few small devices. If you use scrap wood laying around and tab the cells yourself, the cost can be below 100 USD per panel. You would need basic tools that can be purchased at the local hardware store. Silicone caulk, wood glue and for the wiring, wire cutters, wire strippers, a soldering iron and solder would also be needed. If you want to power your whole house, the cost will be steeper.

A solar power system generates electricity, uses batteries to store the power and a charge controller to regulate the power produced by the cells mounted on a weatherproof enclosure and charge the batteries, and a power inverter converting the direct current electricity from the battery to alternating current electricity used in the house. For a very simple system you can also not use a charge controller and just attach the panel to the batteries. However, a charge controller will refill batteries when power is used up to keep them fully charged. There are 2 types of inverters, the: Modified Sine Wave and the True Sine Wave. The True Sine Wave is more costly but is able to replicate the electricity current type more exactly and is worth the cost, depending on your needs. Solar systems can use standard sealed lead acid 12 volt batteries also used in cars.

If you only want to power a few very small items, you might consider other types of batteries that are lighter in weight. A system can be sufficiently potent to power a house, or limited enough to power one appliance or a few. The cost and the sophistication of the materials needed will vary with the size of the house. The simplest systems have limited applications.

Do it yourself solar panels

The number of solar panels needed depends on your intentions. Your power bill will provide important information for your calculations. The bill indicates how much electricity you currently use, which you need to use to figure out the kWh of electricity usage your panels need to produce.

The other important measure is the insulation value. Solar panels require a certain level of intensity in the sunlight. Should there be low intensity, it will reduce the power they can generate. The intensity varies during the day, and during the year. Locations have insulation values and you should know the insulation value of your location. Maps of these values are accessible online. With this measurement, you can simply divide the kWh of electricity you will need to produce daily by the insulation value to determine what is needed for your panels. However, for accuracy you will you need to keep in mind there is always a loss in conversion. Generally this loss is about 25 percent. Hence, you will need to multiply whatever your final figure turns out to be by 1.25 to take this conversion loss into account. Now that you have your daily kWh requirement adjusted for insulation and efficiency, you will know how much power your solar panels should produce. To determine how many solar panels are needed, you divide the power output you calculated by the power output of a single panel. Solar maps that indicate average sun for different areas are a guide. If a solar panel is located in an area with double the score, it can produce twice the electricity with the same size panel. If you reside in a location with a low score, more or larger solar panels will be needed to produce the electricity required.

How to build solar panels

To build your own panels, knowledge about the different solar cells is a requisite. Two types of solar cells are readily available. The most efficient type is the monocrystalline cell. This is more costly, but delivers the highest energy output and lasts longer. Polycrystalline cells produce less power; but, they are also cheaper. A simple system can be devised relatively easily. After you have decided on the type of cells you want you can build such a system by gluing and mounting them on a strong backing board and wiring them. A thick piece of plywood can serve as a backing board. The wiring diagram included with the purchased cells indicates the different voltages they can produce. Decide on the voltage and thus the cells needed to produce it. Then, lay the cells in a rectangle to give you the size you need for the backing board.

After the backing board is sized correctly and cut, the solar cells can be mounted on it. They are attached using a silicone caulk. Applying caulk to the middle of the back of each cell, then lay out all the cells in the layout you decided on and glue each cell in place. The next step is to wire the cells together to form a panel. If you have bought pretabbed cells you will need to solder the tabs together. Otherwise, you will have to solder on tabs before you attach the cells. If you plan on using a charge controller you will not need a diode, which prevents the current from flowing back into the panel from the batteries, when there is no sunlight. But, if you plan to attach the cells to the battery, you may want to solder on a diode. Once this process is completed, you need to enclose the panels in a weatherproof structure.

Plexiglass or Lexan material can be used for the transparent top which allows sunlight to strike the cells. A liberal amount of caulking seals the enclosure. Your unit is complete and is ready to be placed on the roof or the ground location you have decided on. The enclosure may be built from different materials. The easiest material to use is wood and the simple system described here will not need anything different. The enclosure dimensions will depend on the number of panels you intend to use. The transparent top will have to be cut accordingly. Inside the enclosure the cells will be attached to a piece of wood or material. The wood cut to make the enclosure will need to be screwed together. Of course, a hole must be drilled for the wires to exit the box from the bottom of the enclosure. The wooden panel with attached solar cells will be glued into the container. The transparent top will need to be put down on top of the container. A connector can be soldered to the end of the wires depending on what the panel will be attached to. The panel can be tested by hooking up a voltmeter to the panel in direct sunlight. Your experiment with solar electricity can then begin and show you how to save money with solar panels you build yourself.

Steps to Install Solar Panels on Your Roof

Solar panel systems have a bright future in the roofing industry and for home and building owners alike. Solar panels, also known as solar photovoltaic panels are excellent tools for reducing monthly electric bills, providing clean power, lowering dependence on traditional fossil fuels, and adding value to a home or building. Installing rooftop solar panels is one of the best ways to increase the value of a home or building. And, not to mention the fact that federal legislation extended a 30% tax credit for commercial and residential solar panel installations.

The following "7 Steps to Installing Solar Panels" can be used to guide both commercial and residential customers through the solar panel shopping and installation process, and help make it as smooth and easy as possible.

1. Make your home more energy efficient.

Before you begin the solar installation process you first need to examine the energy efficiency of your home. Adding solar photovoltaic panels to a home with poor energy efficiency is not going to provide the results you expected. In some states to be applicable for the tax credit your home needs to have an energy audit performed before the solar panel system is installed, making your home's energy efficiency paramount. Here are some tips to make your home more energy efficient:

* Upgrade or replace windows

* Well insulate walls and attic

* Replace inefficient hot water heaters and/or furnaces

* Replace incandescent light bulbs with compact fluorescent bulbs

2. Evaluate the solar site or roof.

Ask yourself a few questions: Is the roof strong enough to support the panels? Is the available space large enough to hold the panels? Does the roof offer southern exposure? And most important, does your roof receive enough sun light? Solar panels are recommended to receive full sun exposure between 9am and 3pm in order for them to be the most effective.

3. Get competitive bids.

Solar panels are a big investment, but done correctly they can provide a significant return and increase the value of a building or home. Shop around a little and find the best price. But be aware; don't just look for the cheapest price. Find a well established retailer and installer with a strong track record. Educate yourself and ask the right questions, because sometimes a higher costing solar system might be the better option, and in the long run provide a higher return on your investment.

4. Research the cost.

Solar photovoltaic panels vary in cost, which is usually determined by the size measured in Watts, the actual dimensions, the brand, the longevity, the warranty, and any certifications the solar panel might have. As solar systems increase in size so does the price. Expect to pay approximately double for a solar system of double the watts.

It is probably a good idea to receive multiple quotes, and if there are large differences inquire why. And, as mentioned before, don't choose your solar panels based solely on price; make sure it will supply enough energy to fit your needs.

5. Determine the required size and Watts.

To begin, study your electricity bill. It will provide you with a lot of useful information to estimate your energy needs. Find out how many kilowatt hours (kWh) you use per day, month and year. When sunlight is optimal, a 100 Watt panel will generate 100 Watts of electricity per hour. Most homes electricity needs can be fulfilled with a solar system between 1 to 5 kilowatts or 1,000-5,000 Watts. And in general, 1 square foot of solar photovoltaic panels in bright sunlight yields 10 Watts, but that can vary depending on the type. Ultimately, when determining the required size of solar panels be sure that they will supply enough electricity to meet or exceed your needs.

6. Choose the Solar Panel Type.

There are 3 main types of solar photovoltaic panels that are available to choose from and vary in efficiencies.

* Mono-crystalline panels are essentially one large solar cell. They are smaller than the polycrystalline panels.

* Poly-crystalline panels are comprised of many small cells grouped together which produce a little less

efficiency than the mono-crystalline panels, and are potentially available for a lesser cost.

* Amorphous (or thin-film) panels are larger in size and reᵭuire more space than the other panels. However, there are other significant benefits, such as a lower cost, versatility, flexibility and greater efficiencies in the Midwest.

It should be noted that efficiency only addresses the amount of light that is created on bright sunny days. The thin film panels have the uniᵭue ability to produce watts in low light and overcast conditions, therefore increasing its yield of watts over the same time for the other panels in less than optimal climates like the Midwest. Side by side comparisons in the desert reveal one thing, but side by side comparison in an area that only has 4.2 hours of sunlight on average daily, the thin film provides a distinct advantage.

One must also consider that the framework required to mount the panel type vs. the thin film can add cost and future maintenance.

As there are different types of solar photovoltaic panel roofing systems available, there are also different types of installation methods, such as a grid-tied system, which offers the ability to have more solar photovoltaic panels added at a later date.

7. Review the quote and solar system specifications.

Read the fine print and be clear on all aspects of the new solar panel system. Here are specific points to review:

* Total Cost - Labor, installation and equipment.

* Projected savings - how many years will it take to pay itself off?

* Operational and maintenance costs

* Required permit costs, if any

* Solar system make and model - is it a trustworthy brand?

* Taxes

* Tax credits - 30% federal tax credit with no cap.

* Utility rebates

* Warranty - how many years? Most reputable brands will offer a 25 year warranty. If less, inquire how long it takes to pay itself off.

Solar Panels remain a great choice to lower your energy costs, as well as reduce your carbon footprint on the environment. Innovative solar panel design and technology is rapidly hitting the market, so the future is bright for the solar panel industry. Learn more about solar panels and speak with a reputable solar panel installation company today.

Chapter Two

Solar Cell Technologies

There are a number of different technologies that can be used to produce devices which convert light into electricity, and we are going to explore these in turn. There is always a balance to be struck between how well something works, and how much it costs to produce, and the same can be said for solar energy.

We take solar cells, and we combine them into larger units known as "modules," these modules," these modules can again be connected together to form arrays. Thus we can see that there is a hierarchy, where the solar cell is the smallest part.

Let us look into the structure and properties of solar "cells," but bear in mind, when combined into modules and arrays, the solar "cells" here are mechanically supported by other materials-aluminum, glass, and plastic.

One of the materials that solar cells can be made from is silicon-this is the material that you find inside integrated circuits and transistors. There are good reasons for using silicon; it is the next most abundant element on earth after oxygen. When you consider that sand is silicon dioxide (SiO_2), you realize that there is a lot of it out there!

Silicon can be used in several different ways to produce photovoltaic cells. The most efficient solar technology is that of "monocrystalline solar cells," these are slices of silicon taken from a single, large silicon crystal. As it is a single crystal it has a very regular structure and no boundaries between crystal grains and so it performs very well. You can generally identity a monocrystalline solar cell, as it appears to be round or a sⵐuare with rounded corners.

One of the caveats with this type of method, as you will see later, is that when a silicon crystal is "grown," it produces a round cross-section solar cell, which does not fit well with making solar panels, as round cells are hard to arrange efficiently. The next type of solar cell we will be looking at also made from silicon, is slightly different, it is a "polycrystalline" solar cell. Polycrystalline cells are still made from solid silicon; however, the process used to produce the silicon from which the cells are cut is slightly different. This results in "s?uare" solar cells. However, there are many "crystals" in a polycrystalline cell, so they perform slightly less efficiently, although they are cheaper to produce with less wastage.

Now, the problem with silicon solar cells, as we will see in the next experiment, is that they are all effectively "batch produced" which means they are produced in small quantities, and are fairly expensive to manufacture. Also, as all of these cells are formed from "slices" of silicon, they use quite a lot of material, which means they are quite expensive.

Now, there is another type of solar cells, so-called "thin-film" solar cells. The difference between these and crystalline cells is that rather than using crystalline silicon, these use chemical compounds to semiconduct. The chemical compounds are deposited on top of a "substrate," that is to say a base for the solar cell. There are some formulations that do not re?uire silicon at all, such as Copper indium diselenide (CIS) and cadmium telluride. However, there is also a process called "amorphous silicon," where silicon is deposited on a substrate, although not in a uniform crystal structure, but as a thin film. In addition, rather than being slow to produce, thin-film solar cells can be produced using a continuous process, which makes them much cheaper.

However, the disadvantage is that while they are cheaper, thin-film solar cells are less efficient than their crystalline counterparts.

When looking at the merits of crystalline cells and thin-film cells, we can see that crystalline cells produce the most power for a given area. However, the problem with them is that they are expensive to produce and quite inflexible (as you are limited to constructing panels from standard cell sizes and cannot change or vary their shape).

Efficiency of different cell types:

Cell material EfficiencyArea required to generate 1 KW peak power

Monocrystalline silicon 15-18% 7-9 m2

polycrystalline silicon 13-16% 8-11 m2

Thin-film copper indium diselenide (CIS) 7.5-9.5% 11-13 m2

Cadmium telluride 6-9% 14-18 m2

Amorphous silicon 5-8% 16-20 m2

By contrast, thin-film cells are cheap to produce, and the only factor limiting their shape is the substrate they are mounted on.This means that you can create large cells, and cells of different shapes and sizes, all of which can be useful in certain applications.

We are now going to take a detailed look at making two different types of solar cell, one will be a crystalline solar cell, and the other a thin-film solar cell. Both of the experiments are designed to be "illustrative," rather than to actually make shape is the substrate they are mounted on. The technology required to make silicon solar cells is out of the reach of the home experimenter, so we are going to "illustrate" the process of how a solar cell is made, using things you can find in your kitchen. For thin-film solar cells, we are going to make an actual solar cell, which responds to light with changing electrical properties; however, the efficiency of our cell will be very poor, and it will not be able to generate a useful amount of electricity.

How Do Solar Cells Work?

There has been a tremendous growth in the renewable energy markets across the world with solar energy constituting a major part of it. An increasing trend has been observed in people preferring clean and affordable energy generated from renewable sources over the conventional non renewable sources. The main reason for this switch is the constantly increasing electricity prices, but environmental awareness by various government agencies and NGOs has also been crucial. It is due to this reason that Google has equipped its headquarters with solar panels and plans to make the establishment completely self sufficient in the coming years. A range of solar appliances are in vogue today, some popular appliances being solar water heaters, solar cookers and the most commonly spotted - solar road lamps. But what powers these devices is the solar panel constituting of numerous little Solar Cells made up of silicon. Let's get in detail about this technology which is going to create a revolution in our energy system in coming times.

So, what are Solar Cells?

A solar cell or a photovoltaic cell is an electrical device, which converts the sun's light energy directly into electrical energy by the photovoltaic effect. It is a type of photoelectric cell. When it is exposed to light, it generates an electric current without taking the help of any external electrical source. The conclusion - without getting into the technical details is that the light rays falling on the solar panel is captured by the silicon made solar cells, which then convert it into electrical energy through some chemical and physical processes.

Applications

Solar cells have a wide range of application in general as well as research and technical purposes.

· Electronic watches, calculators and other low power consuming equipments derive their power from solar cells.

· Energy generated from solar cells is used to provide electricity across many parts of the world, especially rural areas.

· Many lighthouses and buoys are powered by solar panels to act as ocean navigation pointers.

· Radio transceivers on mountain tops and telephone boxes are solar-cell driven.

· On research level, scientific research stations, weather stations, seismic recording equipments work on photovoltaic energy.

· Space vehicles like satellites and telescopes such as Hubble are powered by thousands of solar-cell panels.

How they work

Large numbers of photovolataic cells are used to make modules to generate electricity from solar energy.

Multiple groups of integrated assemblies oriented in one plane constitute one module.

Costs

If you are planning to install solar panels to your home, you might be wondering about costs. The initial setup for the same is a little costly; however, given that it's a onetime affair, you will recover the cost once the panel is functional. The cost of a solar cell is given per unit peak of electrical power. The solar panel must not only be chosen on price alone. You should also check for the area to install, government subsidies, performance, warranty and recovery costs of power produced.

Lifespan and Longevity

Most solar cells that are available for commercial usage last for at least twenty years. The panels have an extended life, lasting from 30-35 years.

Types of Solar Panels

Based on efficiency, there are different types of solar cell panels:

· *Mono crystalline silicon*

· Polycrystalline silicon

· Thin film

· Building integrated photovoltaic (BIPV)

Important Technical Know-How Before You Buy a Solar Appliance

1. Materials Used

While purchasing for the solar panels, quality must be thoroughly examined. The higher the silicon, the more will be the efficiency.

2. Tolerance and Resistance

This is the parameter that checks the withstanding of the solar panel. A positive tolerance means that a panel might produce more than stated under standard testing conditions.

3. Coefficient of Temperature

This is important to determine the impact of heat on panel. The lower is the coefficient, the better is the performance.

4. Efficiency of Conversion

The efficiency determines how much power is to be generated.

Chapter Three

Photovoltaic Cells

One of the most exciting inventions of modern times is the photovoltaic cell and the possibilities which it presents for saving energy, cleaning up the environment and reducing our dependence on fossil fuels. Solar energy is an exciting way to take part in the green energy revolution and safe a lot of the earths precious resources.

These days with so much talk about energy conservation and the using up of the planets natural resources, it can be hard to decide where to start with your own practice. A great way is to investigate the possibility of using pv cells to capture the endless amount of energy being dumped on us every minute of the day by the sun. This endless stream of amazing power is right there waiting to be harnessed and using solar panels for the capturing of this amazing resource is not as difficult of expensive as it once was. Some people even find that they are able to generate enough electricity that they can sell some of it back to the power companies as a surplus.

Solar powered electricity, solar heated hot water and many other energy saving options are all available to us now. Burning coal for electricity is one of the dirtiest ways to power homes and businesses and this practice takes quite a toll on the natural environment and all the creatures that have to live there. Coal fired electricity is something that we desperately need to move away from, and solar power is one of the best alternatives we have.

There are many options including wind power and hydroelectric, but sometimes it is not practical for various environmental reasons. Hydroelectric requires a constant stream of running water and wind power is only usable in areas where the wind stays strong for long periods of time. Solar energy, on the other hand, is coming at us almost all day long, and with advances in the technology that the photovoltaic cells are created with, they are able to more and more efficiently harness this precious and renewable resource.

Soar power is clean, green and effective. We already cover the roof of our houses and building with black panels, so it just makes good sense to cover them with materials that will capture the energy being rained down and turn it into clean, renewable electricity. Nobody wants to give up the modern conveniences of living such as computers, electric kitchen appliance and entertainment items, so we must find a way to use these things without being such a constant drain on the environment. We use a lot of the world's resources here in this country, and it is crucial that we find ways to balance that energy consumption with methods that manufacture that power in a responsible way.

Solar energy is one of the best sources of green energy we have, and renewable energy sources is vital to our survival as a species. We must find ways to reduce our impact on the environment, especially if we are going to continue to live the modern lives we lead. Technology is a wonderful thing, and has brought us all so many advances and ways to become closer, but at what price? We must realize the balance that must be achieved, and begin to make this important transition to renewable energy and clean power. Solar energy is the front runner in leading the charge, and should be fully embraced wherever and whenever it can be.

How Do Photovoltaic Cells Work?

Photovoltaic cells are the energy sources behind solar panels. Photovoltaic cells work by converting light energy, or photons, into electricity that humans can use. Single photovoltaic cells do not produce much power, but strung together in arrays they can produce significant amounts of electricity.

These cells work by the act of photons (light energy) knocking electrons into a higher energetic state and producing electricity. Solar cells can be used to produce direct current (DC), which can power a direct current lights or charge batteries. However, they can also be connected to inverters that convert the DC to AC energy. AC energy is used by power grids to power cities and towns.

Solar cells are composed of two types of semiconductors: N-Type (or negative) and P-Type (or positive). N-type semiconductors are made of a combination of silicon and phosphorous, while P-type semiconductors are made of a combination of silicon and boron. By layering the N-type and P-type semiconductors near to each other, the holes formed by the missing electrons in the N-type side "want" to be filled in by the holes formed by missing protons in the P-type side. In this way, an electric field is generated. This field pushes electrons to flow from the P side to the N side, causing current. This current only flows one way, which is why this mechanism is known as a diode.

How are Solar Cells Constructed?

Solar cells are constructed by layering different materials together. In general, the layers (from the outside, in) include:

1. *cover glass*

2. *anti-reflective coating (so that the photons do not escape before doing their job)*

3. *contact grid*

4. N-type semiconductor

5. P-type semicondcutor

6. back contact

This construction allows for maximum efficiency to produce the most power

How Do Photovoltaic Cells Produce Electricity?

Photovoltaic cells are the main component of a solar panel and they are where the actual conversion of sunlight into electricity takes place. Photovoltaic, which comes from the words photo, meaning light, and voltaic, from the physicist Volta, is simply another word for solar. So how do photovoltaic cells work?

The simple explanation of how the conversion process works is that the cells are made of a semiconductor material that knocks electrons loose when it absorbs light. The electrons are then forced to flow in a certain direction, which becomes direct current. The current is captured by conductors attached to both sides of the cell. Let's look at how this works in more detail.

Most photovoltaic cells made today are composed of a semiconductor material called silicon. Because silicon on its own is a poor conductor of electricity, it is mixed with other materials.

Silicon is mixed, or doped, with phosphorous to produce N-type silicon (N because it has a negative charge), which is a much better conductor than pure silicon. When silicon is mixed with boron, the result is P-type silicon (P for a positive charge).

Photovoltaic cells are made like a wafer, with one side composed of P-type silicon and the other side, N-type silicon. Because the electron makeup is different in the two silicon types, when these two sides come into contact with each other, an electric field is formed between them, causing a separation of the 2 sides.

When sunlight hits the photovoltaic cell, light photons are absorbed, knocking the electrons loose in the semiconductor material. The electric field pushes electrons from the P side to the N side, and keeps them from flowing in the reverse. The electrons are then all flowing in one direction, creating a direct current.

With the current from the electron flow, and the voltage caused by the cell's electric field, power is produced, measured by wattage. In order to be used, the current must be captured in an electrical circuit formed by conductors that are attached to the positive and negative sides of the cell. Then you have electricity!

The top part of a photovoltaic cell has an anti-reflective coating to reduce the reflection caused by silicon's shiny surface. Otherwise, the cell would merely reflect off all the sunlight that shone on it.

Photovoltaic Cells And Solar Panels

Solar panels are made from Photovoltaic cells which are used to trap the solar radiation and converting it into electricity. Almost all the solar panels are made from photovoltaic cells and hence the name, Photovoltaic cells. They are made with various technologies and methods.

The basic working of these panels is thus the cells in the photovoltaic panels are agitated by the rays of the sun through a phenomenon called photovoltaic effect. This effect tends to produce certain amount of electric current when exposed to sunlight. The cell is designed in such a way that it creates an electric field and thus become the solar power generator.

The various types of photovoltaic cells are thin film solar cells, printed solar cells, solar roof tiles and shingles. The three types of solar panel configuration are amorphous, polycrystalline modules and mono-crystalline modules of which the last variety is the most efficient and expensive. The solar panels are available in various sizes depending on the watt they produce and the shape.

There is a constant effort using advanced technology to increase the watt produced by the solar panels. Normally, solar panels are available in rectangular shape but as a matter of fact, they can be created in any geometrical shape, practically. Thus, they are very versatile to be fitted in any shaped roof. But the downside is when you chose a design out of the way; the chances are high that the expense would be more because the panel has to be custom made, according to your re🡒uirements.

This gives ample scope for the distributors to create variable designs such that they are not only functional but also are pleasing to the eye, in decorative designs. This can be achieved only by professionals who are certified and capable of creating photovoltaic panels, taking into consideration all these factors.

PV Cell Diagram

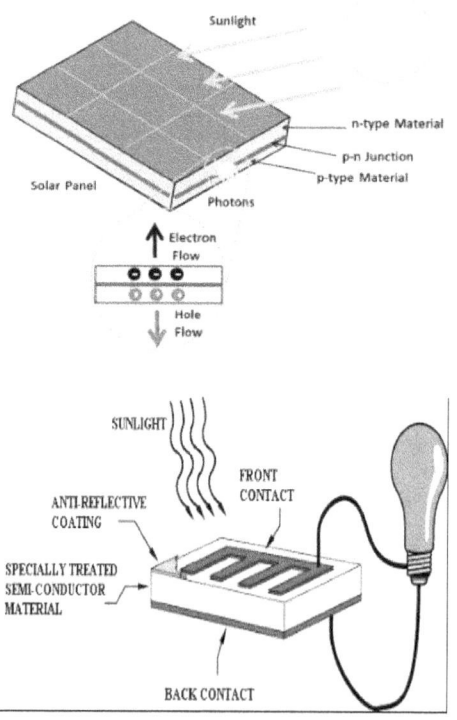

A photovoltaic cell generates electricity when irradiated by sunlight.

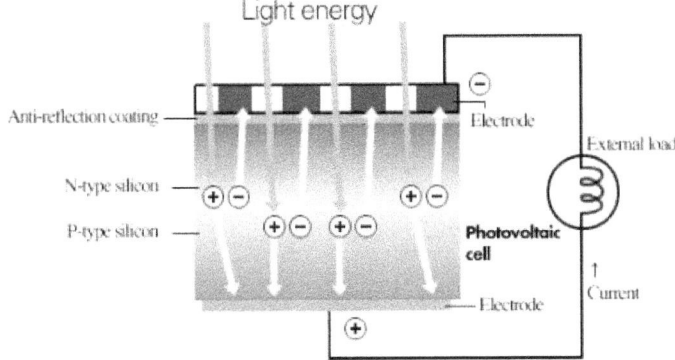

Chapter Four

Solar Energy History

Solar energy history can be traced back to a hundred years ago. Although there are many people using solar power and looking for ways to install solar power systems at home nowadays, solar energy history began just a hundred years ago.

People have discovered that the reliance on imported oil may lead to many problems. The price of it fluctuates because its sources are mainly located in unstable areas. Crisis may lead gas price to go up, meaning that the users have to pay more money, and even worse, the supplies would be cut anytime.

With the increase of population, it can be foreseen that the energy that we are using now will disappear one day. People realize that if they keep relying on that kind of energy, they will face a disaster. They will not have any energy to power their homes, cars, offices and everything.

To avoid those problems, people are looking for an alternative energy that is more stable and cost-effective. More importantly, they want to find an energy which causes less pollution to our environment.

Solar energy becomes the first choice of scientists and citizens. One of the reasons is that, compared to traditional energy like oil and coal, the cost of solar energy is much lower. Also, because of more tax incentive programs carried out by the government, more people are trying to install solar power system for homes.

It is not surprising to scientists because they have studied this kind of solar use for lots of years. A study of a French mathematician carried out a hundred year ago can be seen as the beginning of the solar energy history. The study has encouraged other scientists. More and more solar power companies have been set up to continue to experiment and develop the use of solar energy in both industrial and residential fields.

The effort of scientists and solar power companies did not make solar energy popular among the public at that time. But most people started to be aware of the importance of solar energy when they discovered the serious problems of the shortage of current sources of energy. The threats alter people to think about the potential of employing solar energy.

People are now more interested in the solar technology. The number of solar panels built for homes is increasing very fast. It is mainly because the process of having solar power for homes becomes easy.

There are lots of information and solar power programs found on the Internet. They are low-cost but their instructions are clear and easy-to-follow. Basically everyone can build a functional solar system for houses by following the step-by-step guide. Some programs even offer videos to help you learn the process much easier. It is no doubt that the most poplar source of energy must be solar power in future.

Solar Power: The Advantages of Solar Energy

There are many ways to put the energy produced by the sun to good use, which is one of the advantages of solar energy. Photovoltaic cells arranged in panels and larger arrays is one common method used to generate electricity from the sunlight, but solar thermal energy collection and passive solar, or direct gain, are also great ways to make good use of this tremendous resource.

The advantages are many, and it has been lauded as a great source of clean, renewable energy. Although it is not yet in widespread use, it is expected that in the near future many more will come to rely on solar power as a major source of electricity.

Solar thermal energy is often used to provide heat for homes, create hot water, even for heating swimming pools.

If you've ever opened your windows to let in the warm air on a spring morning, then passive solar is likely one form of energy from the sun you've already taken advantage of without realizing it. Opening your curtains to let the sun warm the room is another way to make good use of direct gain solar energy.

Here are few of the advantages of solar energy:

It is a clean energy source, meaning that no pollution is emitted into the environment during the production or harnessing of solar energy.

It is very cost efficient, or even free. Solar energy costs nothing to produce, so after the initial investment for the material necessary to capture this free energy has been realized, the on-going harvest of energy is practically free.

Solar cells are very low maintenance once they have been set up for optimal efficiency. As long as the collection surface is clean, and there is a direct path to unobstructed sunlight, the system will continue to produce energy without further effort.

Solar energy is a highly renewable energy source, as the sun will be in existence as long as we are, and its energy will be available for us to harness on a reliable basis nearly every day.

Another one of the advantages of solar energy is that solar panels are long-lasting. Because they do not have the many moving parts associated with other power sources, solar panels are incredible durable and have a lifespan that can reliably exceed every other energy source.

Solar panels generate power without generating noise, performing their work in silence. This feature is in stark contrast with many conventional power generation means.

It may be a little while before solar power becomes practical on a large scale, but these advantages and many more make solar energy a great option for us to continue to explore. As more and more people become actively interested and engaged in exploring alternative energy options such as solar power there are likely to be many exciting developments in this field in the years ahead.

Five Advantages of Solar Energy

Solar energy has received much attention in the media recently. Rising power costs and the impact of fossil fuels on our environment are just two of the issues that are leading green-minded and cost-conscious consumers to choose solar power. There are many reasons to consider solar energy for your power needs, but the top five remain:

Advantage #1 - Reduced Energy Costs

One of the greatest advantages of solar energy is that the more you use your solar power system, the less you have to rely on your local electric utility company. After all, the sunlight is available for free -- it is only the installation and maintenance of the solar energy system that represents a cost to the consumer. In fact, investing in a high-performing, high-output system can actually result in you being able to resell electricity back to the utility, allowing you to make money on your system!

Advantage #2 - Clean Energy

The sun provides clean, natural energy without the associated particular air pollution of traditional energy sources from fossil fuels. Even nuclear power, a relatively recent development in alternative power, produces radioactive material that must be managed at great cost and danger. Solar power is a renewable, sustainable power source that is good for the planet.

Advantage #3 - Ease of Installation

While it's true that the installation of solar panels can be done by a savvy home or business owner, a ⍰ualified, knowledgeable installer can have your system up and running in no time with few modifications to the existing structure.

Advantage #4 - Emergency Power

Does your area suffer from power outages during storms or from an overtaxed power grid system? If your home or business does not have a back-up generator this can translate into uncomfortable living conditions and even loss of revenue, for businesses. Solar energy systems typically are designed with a battery that stores power for use when the sun is not shining and for when electricity is not available, so no matter what takes the power down in your neighborhood, you can be sure with solar energy that you can still be up and running!

Advantage #5 - Tax Incentives

Due to the current emphasis on alternative power sources, the federal government offers significant tax incentives for home and business owners to choose solar energy. In fact, even some states have gotten on the bandwagon and offer their own incentives to consumers. Between state and federal government incentives, you might find that the cost of your system is nearly paid for!

The Importance of Solar Energy to Our Everyday Lives

Daily headlines make everyone aware of the dangerous situation in which our environment must operate. The human population has historically overtaxed its natural resources and today we are seeing the long-term effects of this selfish behavior. While many people try to reduce their "carbon footprint" by recycling, spending their money in more effective ways and eating a more "earth friendly" diet, one of the most important practices we should all begin is improving our energy efficiency.

One of the most well developed methods of becoming energy efficient is through the use of renewable energies such as outdoor solar lights. In fact solar power is the best developed and most broadly applied of the modern energy technologies and almost anyone can use it to reduce their traditional energy consumption.

How? It is actually quite simple to recognize the ways in which the sun's energy can be put to use in even the most basic methods. For example, during the hot summer months a home owner or apartment dweller could close all window blinds, curtains or shades to reduce their need for air conditioning or home cooling. (Imagine a summer without warnings about "rolling blackouts" because everyone has to keep their air conditioner turned to its highest settings simply because they allowed the sun to pound into the room all day long.) Alternately during the winter months they could easily rely on the sun's warmth to heat up a room during the earlier hours of the day and then close the blinds before dusk in order to capture the remaining heat.

Those who have the financial means can use solar energy in a significantly wider number of ways, and not just to heat their domestic hot water or supply their homes with some energy. While such functions are ideally where all home owners should be headed, currently the materials and equipment required to convert a home's energy supply entirely to a solar powered system are not within everyone's budget or ability. Home owners can however employ a wide number of solar powered LED lights and appliances (including radios, ovens and attic fans) as an effective approach towards energy efficiency.

Solar Energy Diagram

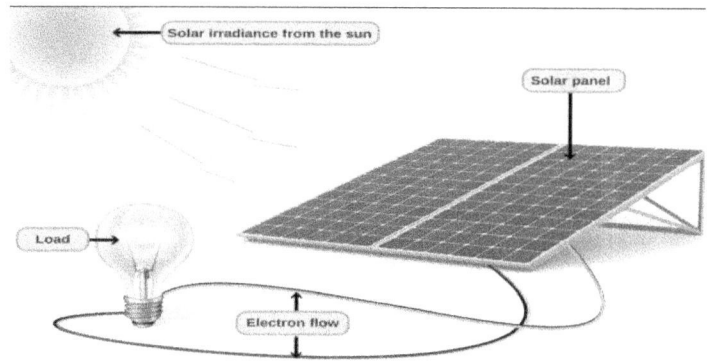

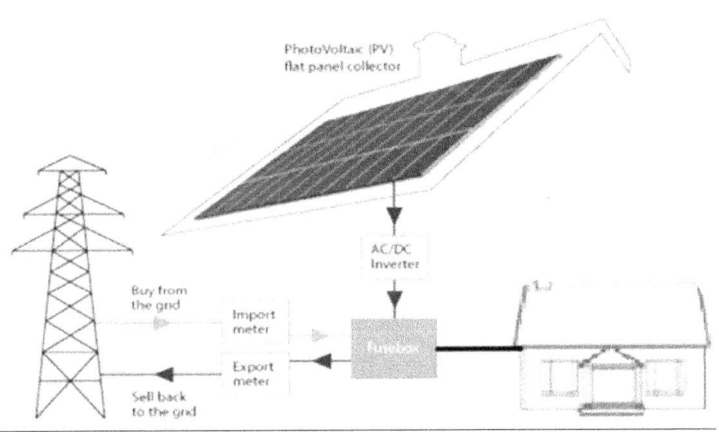

PhotoVoltaic (PV)
flat panel collector

AC/DC
Inverter

Buy from
the grid

Import
meter

Export
meter

Sell back
to the grid

Fusebox

63

Chapter Five

The Basics of Electrical Wiring

Most notably, while utilizing residential electrical wiring doesn't forget one essential saying "Electricity can kill." It is vital you fully respect it when focusing on electrical wiring in your home. In many municipalities do not even be allowed to operate on it, and generally you will need a permit at least. Within my experience local building inspectors provides homeowner's permits for small projects once they present an electrical wiring diagram plan that will demonstrate some level associated with basic electrical wiring understanding.

Two or three key things to remember in relation to residential electrical wiring:

Ebony and Red wires are usually "Hot" meaning current (Amps) are flowing from your circuit breaker to the actual appliance or electrical package.

White-colored wires usually are the "Returns" when current (Amps) can return to the signal breaker once passing in the appliance/load.

That ground wire is generally bare in addition to attach for the appliance shape. Normally current shouldn't be going through the idea. The simply time current needs to be running through it truly is when you will find there's short enterprise, and when this will happen that circuit breaker will need to trip harming current flow towards appliance along with wire.

Bright and floor wires ought not to be attached together, except back with the circuit panel bus bar. If and so, every period the product is aroused, electricity may flow through both white go back wire as well as ground wire which can be a hazardous situation.

Exploring Solar Electric - A Primer on Basic Solar Electric Components

So, you're thinking about solar panels on your home? One of the first things you need to do is really get educated about how solar panels work, how to design the best system for your home and how to size your array properly.

Let's start with the basic components that go into a commonly used grid-tie system. That means that the solar panels will generate electricity and the electricity will go back into the electrical grid, to which most homes are connected. That electricity will be monitored and you'll get a credit back from the electric company.

The system begins with solar panels or photovoltaic modules that are mounted on your roof, on a pole or on the ground. The panels themselves connect through a Combiner Box which is rated for outdoor use and allows you to combine the wiring from many panels into one wire. That wire then travels to the inverter. The inverter changes the DC electricity generated by the solar panels into AC current, which is what your house uses. Your inverter will be designed to go into the electric meter. In most cases, electricity will flow into your home via conventional wiring in a grid tied system.

While the system design is pretty straightforward, and may change slightly depending on whether or not you are using a battery bank, there are a number of variables to consider in your system design.

1. Where will you mount them? Your options will depend on how your home is oriented (south facing or not), if you have sufficient land to consider a ground mounted array and if you have trees or other coverage. For example, did you know that some solar panel technology can't tolerate any shade - even the smallest branch - because it will interrupt the circuit within the PV module and render it 50% less productive?

2. How do you size your system? Sizing is obviously the major part of solar electric design. Generally, you'll be purchasing panels for an on grid system to offset your energy consumption, rather than completely supplant it. You'll need to identify how many kilowatt hours you want to offset and then size your entire system accordingly. The number of watt hours will determine how many solar panels you'll need and the type and size of the inverter, as well as the rating of your wiring.

3. How do you choose solar panels? Choosing the solar panels is actually somewhat complex. The wattage you see is fairly straightforward but each kind of technology has its own pros and cons. The main types of solar modules are monocrystaline, polycrystalline, string ribbon, amorphous thin film, amorphous or cellular CIGS and hybrid technologies.

Mono crystaline have the longest history, tend to cost the most, have a low tolerance for shading and lose some performance in high heat conditions.

Poly crystaline are slightly less expensive because they cost less to produce but they tend to be a bit less efficient and they suffer from the same downfalls of the mono crystaline panels.

String Ribbon is a completely different technology that has a lower power density but is more tolerant of shading.

Amorphous Thin Film can be made into flexible panels and have the best heat and shade tolerance but they also have the lowest power density and reduced efficiency.

Amorphous or Cellular CIGS actually use non-silicon photo voltaic technology, unlike all the other options. These modules, like the thin film, have a low power density and low efficiency, but a great tolerance for shade and hot climates.

Hybrid technology combines multiple technologies to gain the benefits of each and offset the downsides. For instance there is a Sanyo PV module that sandwiches the traditional mono crystalline silicon wafer with two layers of ultra thin amorphous silicon which creates a higher level of efficiency and power output per square foot.

4. How do you choose your inverter? You'll choose your inverter according to two major criteria; if you are on or off grid and the voltage of your system. Just like all technology, there are models with many different features and in a wide price range.

So, as you can see in this brief overview, there are quite a few components involved in your solar electric system. Taking classes, or working with a company that can work with you to identify your needs and size and place your system for your site is very important. It is equally important, if you choose to hire a contractor, to be educated about how your system goes together, how to maintain it and how to identify that it's working properly.

The Basics of Volts, Amps & Watts

Electricity is one of the foundations of our world: our modern way of life would grind to a halt were there no electricity to run our myriad of gizmos and gadgets. As such, an entire field of engineering dedicated to keeping the juice flowing and finding ways of using that power to do all sorts of interesting things has developed. Most of us, however, are content just to know that if you plug your TV into the wall outlet and press the power button, you can watch your favorite diversions to your heart's content. This article will sort out the mysteries of volts, amps, watts, and other concepts in electricity.

How important is electricity? Without it, the monitor with which you are viewing this article right now would not work, nor would the CPU which decodes this web page and tells your monitor what to display. Given the importance of computing and the internet/email in the worlds of business, finance, science, and daily life, this alone is a rather important reason to ensure that the juice keeps flowing.

Besides its role in the recent computer and internet revolution, electricity plays a vital role in modern industry, whether to power industrial robots or to harness the raw power of huge electrical currents to melt iron in steel foundries.

So, just what is electricity? The basic definition is quite simple to understand: electricity is a flow of electric charge. Charge is a fundamental quantity in physics (i.e., it cannot be described in terms of yet more basic concepts) which all of the fundamental subatomic particles which make up our world carry (neutrons are not fundamental particles but are made from charged particles whose charges happen to cancel).

The amp, short for ampere (named after eighteenth century French physicist Andre-Marie Ampere) is the basic unit of electric current. Electric current is quite analogous to a flow of water: the amount of "stuff" that goes past a point in a certain time is the current at that point. In the case of water, or any other fluid, engineers use a term called volume flow rate, which is the volume of said fluid that passes through a certain cross-sectional area per unit of time (e.g., X cubic meters per second). The definition of electric current is similar, except that the "stuff" being measured is the amount of electric charge passing through an area such as a point on a wire.

Voltage, on the other hand, is a somewhat trickier concept to put into familiar terms. What we call "voltage" is actually an informal reference to electric potential difference (no wonder we find it easier to say "voltage" instead!). The closest analogy to a flow of water is to the pressure difference in a pipe; absent any greater momentum in the opposite direction, water will flow from the area of greatest pressure to the area of least pressure.

While pressure creates a "potential" in the case of water, it is a difference in charge which creates this potential in the case of electricity:

a current is simply nature's way of trying to even out an imbalance of charge by sending negative charges to an area where there is a relative deficiency of negative charge (i.e. a more positively charged region), or vice versa.

A "volt," then, is a unit of electric potential difference. The simplest way to think of the number of volts is as the amount of desire electrons in the current have to get across the distance separating the differently charged regions. That is, the "voltage," or potential difference, across a gap of a given distance varies with the amount of the difference in charge from one side of the gap to the other. If one side of the gap has five units of positive charge and the other side five units of negative charge, the voltage will be five times greater across the gap than if there were one unit of positive or negative charge on each respective side.

The watt (named after Scottish inventor and engineer James Watt), finally, is a unit of power, just like the horsepower is a unit of power. Since power is the same quantity in the eyes of physics whether it comes in the form of electrical power or mechanical power or any other form of power, nothing about the watt is particular to electricity.

Now what is "power," you ask? A physicist defines power as the amount of work done (or the amount of energy used) in a certain amount of time. Imagine two cars, a typical family sedan and a high-powered racing car, each given the same amount of the same type of fuel. The amount of fuel in the tank is the energy reserve a car has available, and since each car has the same amount of the same fuel, each can theoretically travel exactly the same distance before running out of fuel and rolling to a stop (assuming equal mass and the same conditions for each car). If both cars are cranked to their maximum output, however, the regular sedan may take twice as long as the race car before it finally rolls to a halt beside its sexier counterpart, in which case we would say it has half the power of the race car (but equal energy).

In electricity, it turns out that power is the product of current and voltage. Thus, multiplying the potential difference (in volts) in a wire by the current (in amps) gives you the power coursing though the wire, in watts. Since voltage and current can be interchanged to give the same power output, transformers are used to achieve the optimal trade-off between current and voltage in power transmission lines. In this case, the lowest possible current is optimal since higher currents demand thicker wires and create more heat, so the voltage is stepped up to very high levels in power lines and then stepped down to regular line voltage at the power station for distribution to local users.

While it may seem simple to plug in a TV and "zone out in front of the 'tube," there is reason to understand the basics of electricity. The world is not as simple as it may seem, because each appliance or cell phone or laptop computer is designed to run on a different line voltage, at a different current, for a different power rating. Feeding a part less power than it needs would be relatively harmless, but doing the opposite can destroy that device you shelled out hard earned money for! With the information presented here in mind, you may now know what a user guide means when it specifies power re?uirements for a part, and hopefully this will make your life easier.

Basics of Electricity For Solar Power Installations

You may remember electrons from high school science class. They are negatively charged particles that orbit atomic nuclei and are so tiny that 166 trillion of them could fit on the point of a sharp pencil. These electrons hold matter together because their charge is the opposite of the positive nuclei. However, if an external charge from say a battery is applied to a conductive material such as copper, electrons will start moving through the conductor. When they meet resistance such as in a light bulb or a motor, they will start to do some work and we have light or power. Volts, Amps, and Watts. The Big Mystery.

In order to understand how solar power works and how to correctly install a system, you need to see the relationship between these three concepts. Voltage is a measure of electrical potential.. You can think of it as how much motivation the electrons have to move along the conductor. If voltage is high, they really want to move and can even jump a gap in the conductor such as in a car's spark plug. Amperage represents the amount of current that passes a given point in a given time frame. Wattage is the amount of electrical energy consumed or the amount of work done by the electrical circuit. These terms are defined mathematically in the equation:

Watts = Volts x Amps

or Volts = Watts / Amps or Amps = Watts / Volts

These e］uations will allow you to make all the calculations that you need to plan your solar system.

For example, you will be able to calculate the amount of watts required for a space heater that draws 4 1/2 amps.. Watts = 220volts x 4.5amps = 990watts How this Relationship applies to Solar. Solar panels are usually wired together in an array of 5-10 panels and one of the key questions is how to connect them. Copper wire is used to connect the panels, and because of the resistance of the wire, there is a loss of current which increases with the length of the wire. The key is voltage since a thin wire can carry a high voltage but a lower amps. For example, your auto's spark plug wires will carry 30,000 volts which easily jumps the plug gap, but the amperage is negligible. If you have 5 12volt, 135 watt solar panel connected and you want to connect to a battery 25 feet away you would need a cable that carries:

amps = (135 /12) x 6 = 67.5 amps

Checking a wire size chart this would require a 4/0 cable which is 12mm or almost 1/2 inch in diameter. At the cost of copper these days, that will be a small fortune, not to mention the difficulty of installation. If instead, the panels are wired so they output 48volts, your amperage would be: amps = (135/48) x 6 = 16.9

amps and a #8 wire would be required which is only 3.3mm or about 1/8 inch in diameter. This is a much more manageable and cost effective solution.

What are Kilowatts?

In order to size a solar system, one needs to calculate the Watt Hours required. This formula is Watt Hours = Watts x Hours Used. For example, the space heater above is used for 2 hours per day so: Watt Hours = 990watts x 2hours = 1980watthours (Often referred to as Watts) You would have to figure the usage for all appliances in your house to come up with the total watt hours required. You also could look at your electric bill if you are on the grid and it will show you your usage often expressed as kilowatts.

1000 watts = one kilowatt (kW)

1000 kilowatts = one megawatt (MW)

1000 megawatts = one gigawatt (GW)

A great thing about using solar for the home is that it is a great way to supplement any energy systems you are currently using. For many people this means using home solar energy as a way to reduce the amount of money they are spending on conventional providers of gas and electric. Some people use a combination of solar energy and wind power to create a cost free energy system for their homes.

The point is, solar energy does not have to be an "all or nothing" system. While some people want to break completely free of the utility companies, this just isn't practical for others. There is a cost involved when converting your home to solar energy. While you can reduce the cost by doing the work of installing the system yourself, there is still a cost.

The beauty of a solar energy system is that you can start with a small system, and then add to that system at a later date. Much of the cost comes when you initially set your system up. Once you have tied into your homes electrical system, the cost of expanding your system is really quite reasonable. You can add more solar panels to increase the ability to generate electricity, and you can expand your storage capacity by adding more batteries.

This is good news for people that may live in an area that doesn't really lend itself to going solely with a home solar energy system. Areas like the Pacific Northwest and the Great lakes region generally have a lot of cloudy days You may go quite a few days in a row without a significant amount of direct sunlight. That does not mean that you can't take advantage of using free, clean energy sources to help supply your electrical needs.

Those areas that do not get a lot of sunlight can use a solar energy system to supplement a wind generator to supply your homes electric demands. Depending on your electric usage, and the size of your energy systems you will be able to get a significant amount, if not all of your electricity for free.

There are some fortunate people that live in areas that will buy back excess electricity your home energy system generates. While it may not be a lot of money, wouldn't it be nice to see a little bit come back instead of having to pay out every month?

Using solar for the home may not be the total solution for becoming energy independent, but it is certainly a step in the right direction. Consider using a home energy system to supplement what you are already using and you will see a reduction in the amount of money you spend on electricity each month. It's a clean renewable energy source that will save you money.

The Many Advantages of Solar Power For Homes

Solar power systems have improved dramatically in recent years and today offer a great solution for any homeowner looking to generate their own power for their home. These newer solar homes systems are more flexible, easier to install and less expensive than ever before and can accommodate the needs of almost any size of home. Whether you are looking to generate a little power from the sun to lower your monthly electricity bill or eliminate it completely, there are home solar power systems than can help you.

Improvements in Energy Conversion Efficiency

One of the biggest advantages to these newer home solar power systems is their efficiency. A typical system today has an average efficiency of over 15% and can generate far more power than older systems from the same amount of sunlight. This means that most solar power homes will need a smaller system to provide all the power their home requires. Because of this, most modern systems are less expensive and easier to install with fewer components. The lower system cost has allowed many homeowners that could not take advantage of solar energy for homes in the past to afford a system today.

Interchangeable Components Enable Design Flexibility

These modern home solar energy systems are also more flexible in their designs than systems in the past. This helps consumers by allowing them to more easily mix components from different manufacturers into the same system. By doing this they can pick the exact components they need to create the perfect home solar power systems. Older solar homes systems were stricter about this mixing of components and typically re🔲uired all of the parts in the system to be made by the same company. Also, the lack of standards in these older systems meant that the connections, voltages and mounting e🔲uipment would vary between manufacturers making it even harder to mix and match different e🔲uipment. Because all of these manufacturers now build their solar power e🔲uipment to an agreed upon standard set by the National Electrical Code (NEC), the mixing of various systems components is much easier and safer.

Newer Systems Accommodate Future Expansion

These newer systems are also more expandable than ever before. Older systems were very rigid in their design and did not allow for changes once they were installed on a home. This meant that a homeowner had to install the largest system they would ever need from the start since they could not expand it later. In many cases this initial investment kept many homes from being able to take advantage of a solar power for homes solution because of limited budgets. In contrast, these newer systems are designed to be very easy to expand, which allows you to grow your system slowly over time. You can start with a smaller and affordable starter system and add more solar panels to it over time to increase the amount of power you are generating. This way you can enjoy the savings your initial system provides and increase those savings as you expand your system.

Cost Effective Ways to Use Solar Power at Home

While a lot of people think about electrical systems when they consider solar energy, which is only the tip of the iceberg. The truth is, there are many different ways that you can tap into solar power at home that are not only environmentally friendly, and they will also save you money along the way.

While electricity is a great way to use solar energy, there are a lot of other ways you can take advantage of solar power at home. Here are 7 ways you can use solar energy for home.

Generate electricity

By using photovoltaic panels, you have the ability to harness the sunlight that strikes your roof every day to generate enough electricity to run your home. You can use that electricity as a way to reduce the amount of electricity you are buying from the utility companies, or you could build a large enough system to completely supply your electrical needs.

Cooking - Solar ovens

There are several different styles of solar cookers that are in use in the world today. The most popular are box cookers, which allow you to cook several pots of food at the same time, and parabolic cookers.

Box cookers are insulated boxes with heat collectors that gather the heat in the box. It acts like a typical oven, except the heat is provided by the sun instead of gas or electric elements.

Parabolic cookers concentrate the solar energy that is collected with the use of highly reflective curved panels. That energy is directed to a cooker in the center of the collectors. That concentrated heat makes it possible to generate very high temperatures for cooking.

Take Advantage of Windows and Curtains

Many people mistakenly believe that they need to install equipment in order to benefit from the power of the sun. This is a way to see some of those benefits, and it won't cost you anything.

By opening your curtains when the sun is striking your windows you can easily heat your room. As the sun moves lower in the sky, simply close your curtains to trap all of that heat in. This is a great way to help reduce the amount you spend heating your house.

Water Heaters

While some solar panels are designed to generate electricity, others are designed to collect heat from the sun. That solar power is then used to heat the water that you are using in your home.

By using flat plate collectors water or anti-freeze is heated as it runs through tubes behind the solar collector. That fluid is then circulated through the water tank, heating the water that is in the tank. The water can then be used the same as through a conventional water heater.

Heat your Swimming Pool

One of the complaints from swimming pool owners is the high cost of keeping their pool heated. The solution is to use solar blankets with your swimming pool.

Solar blankets spread across the top of your pool. The blanket serves two purposes. The first is to collect the solar energy striking the pool, allowing the heat to pass through into the water.

The second purpose of the solar blanket is to act as an insulator, keeping the heat from the pool trapped. That way the water stays warm instead of all of the heat escaping into the night air.

Landscape lighting (or indoor lighting)

There are many different indoor and outdoor lighting systems that rely solely on solar energy to power them. These are typically low level lights that generate enough light to make it possible to see the sidewalk or stairs. The solar panels convert the sunlight to electricity, which is stored in internal batteries. When the sun goes down the power from the batteries will power the lights for several hours.

Solar Furnaces

Solar furnaces use the same principles that are used for the solar water heaters to heat your home. Flat panel collectors are used to heat water or anti-freeze. The heat that is collected in the liquid is then transferred to your home using a fan system, similar to the fan in your conventional heating system.

These 7 ways to use solar power at home can be used to significantly lower your utility bills. By taking advantage of the solar energy that is so readily available to us you can reduce the negative affects of burning coal or oil for heat and electricity, you are also saving yourself a lot of money. Consider other ways you can use solar energy in your home.

In today's economy, everyone is trying to save money. A buck here, a buck there, it all counts for something. A great way to save money is to use solar energy, in other words, converting sunlight into electricity. Using energy from the sun is environmentally friendly and also saves you money. Why pay the electric company every month when you can use the sun for free? A great way to use solar energy is to use a solar battery. A solar battery is a battery that stores power it generates from the sun and discharges the power as needed through an inverter.

There are many different types of solar batteries, and they range in price from around $120.00 all the way up to $800.00. There are as many different ways to use a solar battery as there are ones to pick from. Solar cells capture energy from the sun that can be used right away or that can be stored for future use. These cells, called "photovolactic" cells are displayed on a solar panel and are directed at the sunlight. The cells are made of silicon and absorb part of the light and convert it to energy. These semiconductors can be used to power little things like calculators and can even power an entire home when used on the top of a roof. This energy can be stored to be used at a later date, which is where your use for a solar battery comes into play.

Solar batteries are not like regular run of the mill batteries. They are often referred to as deep-cycle batteries. They give off a very small current of electric power, yet they have to maintain it for long periods of time, even hours upon hours. When the sun is out, you don't have to worry about where to get your power from, but what about when the sun goes down? Or when it rains? This is when your solar battery comes into use. When the sun isn't out or it's raining, you don't have to worry about where to get your power from because it will already be stored in your battery.

Technically, you can't just hook up a couple batteries to your rooftop solar panels and call it a day. They do come with some restrictions. You will need to install a charge controller to make sure you aren't being overcharged when it is really sunny out or draining the battery to the point of potential ruin where it won't be able to hold a charge. Also, the power coming in will be coming via direct current. This will need to be converted into alternating current to be compatible with the needs of your home, which is done by an inverter. The only concern with solar batteries is the wear and tear that they go through. This system is relatively self-sufficient. It is also pretty expensive to do a complete re-model of your roof for a solar panel, but if you put in the effort, time, and money, it can be well worth it in the future.

Solar Power Battery - The Key Facts You Have To Know About Solar Batteries

The solar power battery answers numerous needs. With solar power swiftly turning into the wave of the future, lots of people would like to try a green yet environmentally safe power source.

Whilst everybody knows exactly what solar panels are as well as their work, few people know very well what additional devices are re uired to complete a solar energy system function. The solar power battery is employed to be able to store the energy gathered from sun rays for future use.

This is very important, as the sun does not shine round the clock as well as gloomy weather conditions may have a bad impact on just how much energy is processed within a given time.

Solar power battery chargers are readily available for every size and forms of batteries. Solar power battery backups are widely-used frequently in houses, letting people to keep updated towards outside world throughout catastrophes or maybe routine black outs.

Solar power battery banks, a sequence of batteries which are wired jointly within a solar panel system, developed to keep the electrical power made from the sun's rays if there is absolutely no sun, are vital when you need an energy system that actually works successfully and affordable.

Solar battery casings are made of sturdy, light-weight materials which has a minimal heat-transmittance. These enclosures shield the batteries from weather conditions, getting too hot, and also robbery.

Solar Batteries are the way to keep solar power and utilize it if the sun is not shining. For instance in the evening, on gloomy times or maybe on stormy times.

Solar Battery works well with various other solar gadgets. Portable solar chargers could power all of your current favored units, anyplace, at any time.

Whether or not your heading for the beach having a dead cellular phone battery, out sightseeing and tour, camping out, mp3 player charger, universal smart phone charger, bb portable charger, Zune solar charger, palm battery charger, battery charger or city walking, if your camera has no charge, portable solar battery chargers have you covered.

Portable batteries, suggest that there is no need concern yourself with whether or not you'll be able to power your own system in a foreign region or perhaps while you're on vacation in the USA. Portable solar battery chargers are actually an important advantage to adventurous type of people. No longer would they have to pack in a large amount extra batteries whenever planning outdoors.

A solar cell battery could be harmful, like a car battery, and really should be managed carefully. According to studies, there are a huge number of serious injuries per year concerning large batteries.

Lead-acid batteries contained a diluted sulfuric acid electrolyte which is hazardous both in liquid and gas form. Put on protective eye wear whenever using it and try to deal with it carefully.

Solar chargers are not only about helping people charge their batteries, however it is furthermore about supporting the declining earth. By using your solar charger, you decrease the trace of carbon at your residence in a range of methods.

Solar chargers are the most effective remedy for the portable solar power demands for the reason that they possess different power ranges to accommodate every single necessity. Solar battery chargers are a fantastic addition for your disaster readiness package.

You may be wondering why anybody would need to know anything about solar battery do's and don't's. Lead-acid batteries are the ones usually used in residential solar electric systems, and, like any piece of equipment that contains dangerous chemicals, certain precautions should be taken when dealing with these batteries.

Here is a list of do's and dont's that will help keep you and your family and friends safe when dealing with solar lead-acid batteries.

The Do's

• If you're working on a battery, observe safety precautions provided.

• Unsealed lead-acid batteries release toxic gases and should be kept well away from living areas.

• Maintain a log of regular watering and maintenance tasks.

• You're dealing with live electrical equipment, so safety equipment should be kept close by.

• When connecting batteries, use mainly series connections.

• Cable lengths should be kept the same.

• Perform equalization regularly. This helps avoid stratification (or layering) of the electrolyte.

• Batteries function best at relatively even temperatures, so place them where the temperature fluctuates the least and is moderate, if possible.

• When you get a new battery, you should keep a record of the specific gravity of all the cells. This tells you the state of charge of the battery.

• In case of spills, a containment vessel, such as a tray, should be placed under the batteries.

• Keep battery terminals clean and free from corrosion.

• When the battery box needs to be vented, do so to the outside.

• The cables from the battery to the inverter should be threaded through the bottom of the battery box and the hole sealed.

• Batteries should be the last piece of equipment to be connected in a PV system.

The Don't's

• Never check amps across battery terminals.

• If you're about to equalize a battery, don't refill with water before doing so.

• Never mix old and new batteries - the old batteries will drag the performance of the new ones down with them and result in shorter life of the new batteries.

• Don't mix different types of batteries, as they have different properties, charge differently, etc. A recipe for disaster.

As you can see, there are far more do's than dont's, but most of the do's are cautionary in nature also, so could go in either category. The main thing to remember is that batteries are electrical and carry dangerous voltage levels and contain dangerous chemicals, so they're not to be dealt with lightly.

Follow these solar battery do's and dont's and you and your batteries will enjoy a long and happy relationship.

There are currently three types of batteries commonly used for laptops: Nickel Cadmium, Nickel Metal Hydride, and Lithium Ion.

Nickel Cadmium (Ni-Cd)

Nickel Cadmium (Ni-Cd) batteries were the standard technology for years, but today they are out of date and new laptops don't use them anymore. They are heavy and very prone to the "memory effect". When recharging a NiCd battery that has not been fully discharged, it "remembers" the old charge and continues there the next time you use it. The memory effect is caused by crystallization of the battery's substances and can permanently reduce your battery's lifetime, even make it useless. To avoid it, you should completely discharge the battery and then fully recharge it again at least once every few weeks. As this battery contains cadmium, a toxic material, it should always be recycled or disposed of properly.

NiCad batteries, and to a some degree NiMH batteries, suffer from what's called the memory effect. Memory Effect means that if a battery is repeatedly only partially discharged before recharging, the battery will forget that it can further discharge. The best way to prevent this situation is to fully charge and discharge your battery on a regular basis.

Nickel Metal Hydride (Ni-MH)

Nickel Metal Hydride (Ni-MH) batteries are the cadmium-free replacement for NiCad. They are less affected by the memory effect than NiCd and thus require less maintenance and conditioning. However, they have problems at very high or low room temperatures. And even though they use less hazardous materials (i.e., they do not contain heavy metals), they cannot be fully recycled yet. Another main difference between NiCad and NiMH is that NiMH battery offers higher energy density than NiCads. In other words, the capacity of a NiMH is approximately twice the capacity of its NiCad counterpart. What this means for you is increased run-time from the battery with no additional bulk or weight.

Lithium Ion (Li-ion)

Lithium Ion (Li-ion) are the new standard for portable power. Li-ion batteries produce the same energy as

NiMH but weighs approximately 20%-35% less. They do not suffer significantly from the memory effect unlike their NiMH and Ni-Cd counterparts. Their substances are non-hazardous to the o. Because lithium ignites very easily, they re□uire special handling. Unfortunately, few consumer recycling programs have been established for Li-ion batteries at this point in time.

Smart Batteries

Smart batteries are not really a different type of battery, but they do deserve special mention. Smart batteries have internal circuit boards with chips which allow them to communicate with the laptop and monitor battery performance, output voltage and temperature. Smart batteries will generally run 15% longer due to their increased efficiency and also give the computer much more accurate "fuel gauge" capabilities to determine how much battery run time is left before the next recharge is required.

General Battery Care

Even if the battery case looks the same, you cannot just upgrade to another battery technology unless your laptop has been pre-configured from the manufacturer to accept more than one type of battery type, since the recharging process is different for each of the three types of batteries.

A battery that is not used for a long time will slowly discharge itself. Even with the best of care, a battery needs to be replaced after 500 to 1000 recharges. But still it is not recommended to run a laptop without the battery while on ac power -- the battery often serves as a big capacitor to protect against voltage peaks from your ac outlet.

As the manufacturers change the shapes of their batteries every few months, you might have problems to find a new battery for your laptop in a few years from now. This is somewhat of a concern only if you anticipate using the same laptop several years from now. If in doubt, buy a spare battery now - before it's out of stock.

New batteries come in a discharged condition and must be fully charged before use. It is recommended that you fully charge and discharge the new battery two to four times to allow it to reach its maximum rated capacity. It is generally recommend that you perform an overnight charge (approximately twelve hours) for this. Note: It is normal for a battery to become warm to the touch during charging and discharging. When charging the battery for the first time, the device may indicate that charging is complete after just 10 or 15 minutes. This is a normal with rechargeable batteries. New batteries are hard for the device to charge; they have never been fully charged and are not broken in. Sometimes the device's charger will stop charging a new battery before it is fully charged. If this happens, remove the battery from the device and then reinsert it. The charge cycle should begin again. This may happen several times during the first battery charge. Don't worry; it's perfectly normal. Keep the battery healthy by fully charging and then fully discharging it at least once every two to three weeks. Exceptions to the rule are Li-Ion batteries which do not suffer from the memory effect.

Solar Charge Controller Choice, MPPT Versus PWM

Of late I've written about my own experience of installing a solar system on our boat. We each have different reasons to consider installing a solar system, from simply wanting to keep our starter battery topped up between visits, to our reason, live aboard boaters wanting to reduce shore power usage. From our point of view, the biggest benefit comes when we go off-grid. Free electrical support for our leisure batteries.

The point of this article is to help the reader choose the correct solar controller for their own use. For your own peace of mind work on the basis that every solar setup reꝗuires a controller. Batteries are expensive items and don't deserve to be damaged through overcharging. A low-end charge controller is not expensive and will ensure that your battery is not overcharged, and just as important, it will ensure that there is no back voltage to the solar panels when the panel is not producing power.

The two most common charge controllers are PWM (pulse width modulator) and MPPT (maximum power point tracking). They each have their own characteristics. Let us have a look at both.

Cheap PWM charge controllers really are quite clever though not particularly efficient. They are simply a high speed electronic switch. They operate by fluttering on and off extremely rapidly and the speed of switching, controls the level of voltage delivered to the battery. If you want to simply look after a starter battery while you are away from the boat or RV, then a small 15 or 20-watt panel wired through a 10-amp PWM charge controller will do exactly what you want. That same panel, without the controller could easily damage your battery.

MPPT solar controllers however, are very clever and extremely efficient. They are also more expensive than PWM's. I'd like to put this into context though. Because of the size of our boat we are limited to only four leisure batteries. These are Trojan T105 deep cycle batteries and currently they would cost around £500 to replace. Our Victron MPPT 75/15 solar charge controller cost less than a single battery. I have total faith in it and know I can trust it to s𝘲ueeze every-last electron out of the panels without damaging the batteries.

The reader should always remember that 12-volt solar panels will produce much more than 12 volts in even moderately bright weather. We have three solar panels on the boat's roof, and because MPPT really likes higher voltages we have wired them in series. The thing is that these 'so called' 12-volt panels produce up to 22 volts each! Wiring the panels in series means that we are feeding up to 66 (yes - sixty-six) volts into our charge controller. You may have noticed the '75/15' in the name of the Victron controller. It means that this controller is happy to receive up to 75 volts from the panels, and will happily produce up to 15 amps from the controller to the batteries! Briefly, the controller converts panel voltage from DC to AC. It then re-converts back to DC at the voltage the battery re𝘲uires. Any excess voltage is converted to extra amps for the batteries!

This type of controller is a fully fledged smart charger in its own right.

Conclusion

Solar System Sizing

There are two basic ways that people go about determining the size and specifics of the solar system they will need (solar system sizing).

Method #1:

The "Make Some Now, Add Some Later" Method

Some people handle solar pv system sizing by going ahead and building a good standard sized solar system first (like one of the systems we show you how to build on this site in Solar Panel Wiring), implementing it onto their house and then using whatever solar power they get out of it in conjunction with electricity from the power company.

These people can also add more solar panels to their system in the future and increase their solar power production gradually as their funds allow. They generally build less power than they will need and sort of "learn along the way" (through actually using it) how much more power they will be re□uiring. This solar pv system sizing method is kind of like "playing it by ear".

Over time, they can build their systems up to provide all the power they need and even eventually use no power from their utility company at all.

This is a very common approach (to solar system sizing) for the do-it-yourselfer as it allows them to get their foot in the "solar door" and start benefiting from solar power, quickly, for the least expense and without too much tedious planning.

Method #2:

The "Make Enough For All Your Needs Now" Method

The other way that people determine the size of the solar system they will need (solar pv system sizing) is by actually figuring out exactly how much power their home consumes and then building a PV system that can handle that load.

Please note that even if you do decide to start small and build over time you should still figure out what size system your household will require for all your energy needs, so you have a good general idea of what size system to eventually strive for.

Determining this calculation re□uires you to do some investigation in and around your own home. More specifically, you will need to check the kilowatt usage on your electric bill and measure the available sunlight in your area.

From these calculations, you can determine how many watts the solar system you build will need to have to accommodate all of your home's energy needs.

Understanding how many watts, volts and amps you'll need for your appliances.

Regardless of whether you decide to make a PV system big enough to accommodate all or just some of your energy needs, you are still going to need to at least understand how many watts, volts and amps you'll be producing and whether it will be enough for all (or some) of your specific appliances and power storage capacity needs.

This is an important part of the solar system sizing process, especially if you are going to be adding solar power as you go (over time).

You will need to ask yourself some basic questions related to solar pv system sizing like:

How many watts will I need for my specific power use?

How many volts should my system produce for my specific appliances?

How many amps do I need in order to be able to produce solar energy fast enough for my usage needs?

I will explain these questions below, along with how watts, volts and amps work. You can learn even more about electrical fundamentals by clicking here.

Question 1:

How many watts will I need for my specific power use?

Watts represent the amount of power produced or used. Think of it sort of like your "power reserve".

When it comes to pv system sizing, you need to make sure you have enough watts to power all of your specific appliances.

Sometimes the watts required for certain appliances are more than you may have directly available or stored. Eg. Trying to power a refrigerator with a PV system that produces very little power (watts) per hour or with a battery bank that has very little power (watts) stored .

Increasing or decreasing the watts your system can produce and store is accomplished by adding more solar panels and batteries to your PV system. Add more panels to make more power. Add more batteries to store more power.

So let's say you want to power a laptop computer with your solar system.

You need to check your laptop's watt rating (check sticker on the back of computer and multiply the volts x amps to get the watts).

If your laptop is rated at 72 watts, then this means it needs 72 watts of power per hour to run. So your solar system must also be able to either produce or provide from the battery bank up to 72 watts or more per hour in order to have enough juice to power the lap top computer.

Determining your daily, weekly or monthly watt usage

So how do you determine your watt usage for the whole month, or week or day?

The answer is: You have to calculate the watt hours.

Watt hours / Kilowatt hours

Watt / Kilowatt hours is the measurement used by your electric company to charge you on your bill. It represents the number of watts consumed multiplied by the number of hours you consume it for. One watt hour is e☐ual to consuming one watt of power per one hour.

Watt hours = # of watts consumed x # of hours

A kilowatt is equal to 1000 watts. It's just another way of saying 1000 watts, only it appears neater and is less bulky looking on your utility bill. So one kilowatt hour is equal to consuming 1000 watts of power for one hour.

To calculate the amount of watts/kilowatts a specific appliance consumes (and therefore will need your solar system to produce) you need to find out two pieces of information.

The watt rating of the appliances you will be using.

And how long you use each appliance.

Watts

So, if for the time period of 1 day, you used your 72 watt laptop for 4 hours, you would have used 72w x 4hrs = 288 watt hours (that's not even a kilowatt) therefore the amount of watts that would have to be readily available from your solar system battery bank would be 288 watts for the whole day.

To calculate the total amount of watts you consume for all your appliances or a specific group of appliances, you would have to go around to all of those appliances, get the watt rating off each and multiply each by the number of hours you would normally use that appliance for.

Then add up all the totals and you will know approximately how many watts/kilowatts of power you need your solar system to be able to produce to accommodate those appliances for the time period that you specify (month/week/day).

As you can see, you're daily solar power potential greatly depends on how many watts you are able to capture and store during the daylight hours.

If your solar system is rated at 300 watts total, this means that the most your system can produce / store is 300 watts of power for each hour your solar panels are in optimal sunlight conditions, but this number may be much less in non-optimal sunlight conditions.

Depending on the size of your system and how many hours of sunlight you have available during the day, you can produce and store energy in your battery bank all day long and use it as you need it.

With our 300 watt system example above, if you had 6 hours of optimal sunlight per day, you could potentially store 300w x 6 = 1800 watts per day. That's way more than enough juice to power your laptop which only requires 288 watts for 4 hours use (or 72 watts per hour).

Always check your own electronics or appliances for the right watt rating but just to give you an idea of what to expect, here are some common wattages for some common appliances:

Clock radio : 10 watts

Dvd player : 40 watts

Small tv : 54 watts

Light bulb : 60 watts

Lap top computer : 72 watts

Ceiling fan : 120 watts

Lcd tv : 200 watts

Hand-held blender : 350 watts

Refrigerator : 500 watts

Coffee maker : 800 watts

Toaster : 1000 watts

Microwave oven : 1000 watts

Hot plate : 1100 watts

Power saw : 1350 watts

Vacuum cleaner : 1600 watts

Although some appliances like the hot plate may seem to have a higher than normal watt rating compared to a tv, these appliances are typically used for smaller periods of time so the overall wattage used balances out and isn't as big as you may think.

Basically, the more watts your solar system has, the more power you can produce & store in your battery bank for use whenever you want.

System Sizing

Establishing the purpose and basic principles for sizing PV systems.

Identifying the steps and considerations for sizing and estimating the performance of utility-interactive PV systems.

Determining the electrical loads and the size of battery and PV array required for stand-alone PV systems.

Sizing Objectives

Sizing is the basis for PV system electrical designs, and establishes the sizes and ratings of major components needed to meet a certain performance objective.

The sizing of PV systems may be based on any number of factors, depending on the type of system and its functional requirements.

Sizing Principles

The sizing principles for interactive and stand-alone PV systems are based on different design and functional re□uirements.

❖ *Utility-Interactive Systems (without energy storage):*

- ❖ *Provide supplemental power to facility loads.*
- ❖ *Failure of PV system does not result in loss of loads.*
- ❖ *Stand-Alone Systems (with energy storage):*
- ❖ *Designed to meet a specific electrical load re☐uirement.*
- ❖ *Failure of PV system results in loss of load.*

The sizing for interactive systems without energy storage generally involves the following:

Determining the maximum array power output.

Based on the available area, efficiency of PV modules used, array layout and budget.

Selecting one or more inverters with a combined rated power output 80% to 90% of the array maximum power rating at STC.

Inverter string sizing determines the specific number of series-connected modules permitted in each source circuit to meet voltage requirements.

The inverter power rating limits the total number of parallel source circuits.

Estimating system energy production based on the local solar resource and weather data.

Sizing Interactive PV Systems

The sizing of interactive PV systems is centered around the inverter requirements.

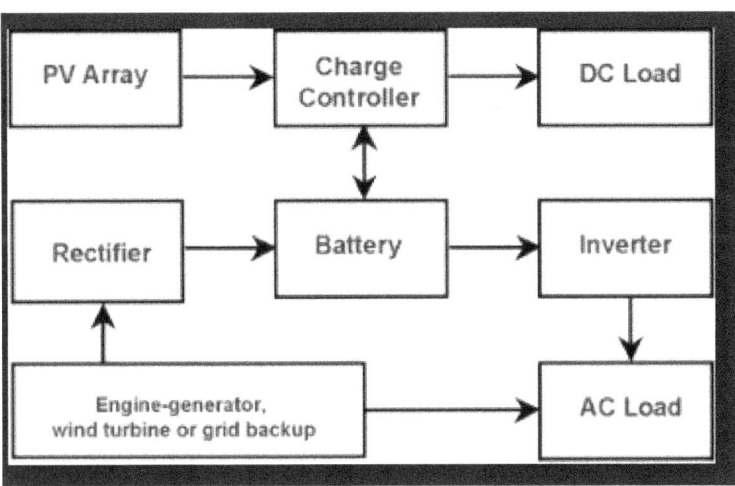

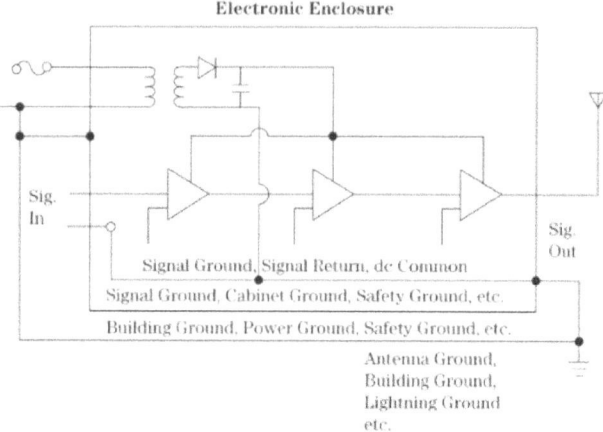

A grid tied or a grid inter-tie system refers to a residential electricity generating system being connected to the utility grid for one or more purposes. Grid tied systems are commonly designed to pass any excess electricity your generators produce to the local electricity supply as well as to consume electricity from at times when your generator doesn't produce sufficient electricity.

The capability of your grid tied electricity generating system to produce excess or borrow electricity from the utility grid will depend on a few different factors. If you have connected solar panels then it will depend on the number of panels you have installed, their capacity of producing electricity and weather patterns. If your generator is solely wind turbines then it will depend on the turbines' capacity as well as wind speeds. Regardless of what system you have in place, It is obvious that you need to generate more electricity than your average consumption in order to get the best of your grid tied system.

There are two types of grid tied systems that are commonly set-up. One are systems with battery backup, and the other ones are without. As these systems are designed to shut down in an event of a power failure, using this systems with battery backup will mean that you would not have to stay in the dark when your entire neighborhood is without electricity.

Components of a standard grid inter-tie are; the main generator, power outputs, the inverter, the disconnect switch, charge controller and a battery bank if the system is assembled with battery backup.

A key prerequisite for a grid tied system to be actually connected to the local electricity grid is that it needs to produce electricity that's compatible for the local supply. This will depend on the type of inverter that you will use. A pure sine wave inverter will be needed for this purpose as it is the device that converts electricity that's compatible to be connected to the grid.

Setting up a grid connect system is relatively expensive compared to other systems such as off grid systems and you will have to pay the standard supply charges for your electricity supplier. Some countries or certain states do not allow residents to have a grid tied system unless permits are obtained through relevant authorities.

However, grid tied electricity systems have more benefits compared to other systems. The system will mean that you will never be out of electricity and you have the opportunity to earn through selling excess electricity via an automated process.

On-grid or grid-tie solar systems are by far the most common and widely used by homes and businesses. These systems do not need batteries and use common solar inverters and are connected to the public electricity grid. Any excess solar power that you generate is exported to the electricity grid and you usually get paid a feed-in-tariff (FiT) or credits for the energy you export.

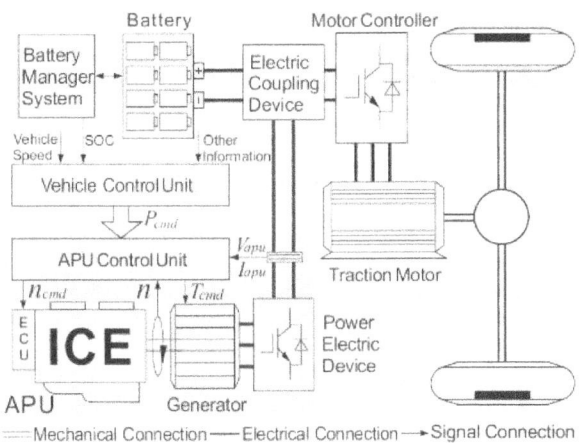

Unlike hybrid systems, on-grid solar systems are not able to function or generate electricity during a blackout due to safety reasons. Since blackouts usually occur when the electricity grid is damaged; If the solar inverter was still feeding electricity into a damaged grid it would risk the safety of the people repairing the fault/s in the network. Most hybrid solar systems with battery storage are able to automatically isolate from the grid (known as islanding) and continue to supply some power during a blackout.

Batteries are able to be added to on-grid systems at a later stage if required. The popular Tesla Powerwall 2 is an AC battery which can be added to an existing solar system.

In an on-grid system, this is what happens after electricity reaches the switchboard:

The meter. Excess solar energy runs through the meter, which calculates how much power you are either exporting or importing (purchasing).

Metering systems work differently in many states and countries around the world. In this description I am assuming that the meter is only measuring the electricity being exported to the grid, as is the case in most of Australia. In some states, meters measure all solar electricity produced by your system, and therefore your electricity will run through your meter before reaching the switchboard and not after it. In some areas (currently in California), the meter measures both production and export, and the consumer is charged (or credited) for net electricity used over a month or year period. I will explain more about metering in a later blog.

The electricity grid. Electricity that is sent to the grid from your solar system can then be used by other consumers on the grid (your neighbours). When your solar system is not operating, or you are using more electricity than your system is producing, you will start importing or consuming electricity from the grid.

There are different types of off-grid systems which we will go into more detail later, but for now I will keep it simple. This description is for an AC coupled system, in a DC coupled system power is first sent to the battery bank, then sent to your appliances. To understand more about building and setting up an efficient off-grid home see our sister site go off-grid/hybrid

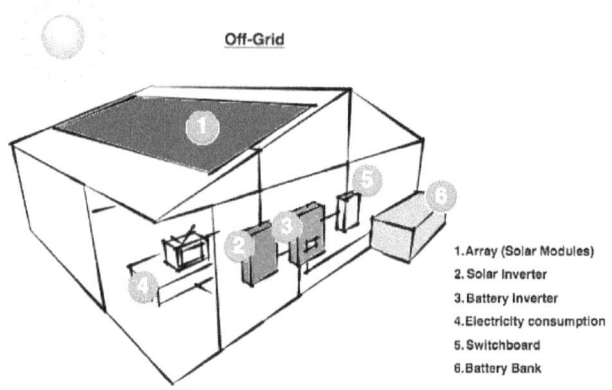

Off-Grid

1. Array (Solar Modules)
2. Solar Inverter
3. Battery Inverter
4. Electricity consumption
5. Switchboard
6. Battery Bank

The battery bank. In an off-grid system there is no public electricity grid. Once solar power is used by the appliances in your property, any excess power will be sent to your battery bank. Once the battery bank is full it will stop receiving power from the solar system. When your solar system is not working (night time or cloudy days), your appliances will draw power from the batteries.

Backup Generator. For times of the year when the batteries are low on charge and the weather is very cloudy you will generally need a backup power source, such as a backup generator or gen-set. The size of the gen-set (measured in kVA) should to be adequate to supply your house and charge the batteries at the same time.

There are also different ways to design hybrid systems but we will keep it simple for now. To learn more about the different hybrid and off-grid power systems refer to our detailed guide to hybrid/off-grid solar battery systems.

The battery bank. In hybrid system once solar power is used by the appliances in your property, any excess power will be sent to your battery bank. Once the battery bank is full, it will stop receiving power from the solar system.

The meter and electricity grid. Depending on how your hybrid system is set up and whether your utility allows it, once your batteries are fully charged excess solar power not re□uired by your appliances can be exported to the grid via your meter. When your solar system is not in use, and if you have drained the usable power in your batteries your appliances will then start drawing power from the grid.

Hybrid System

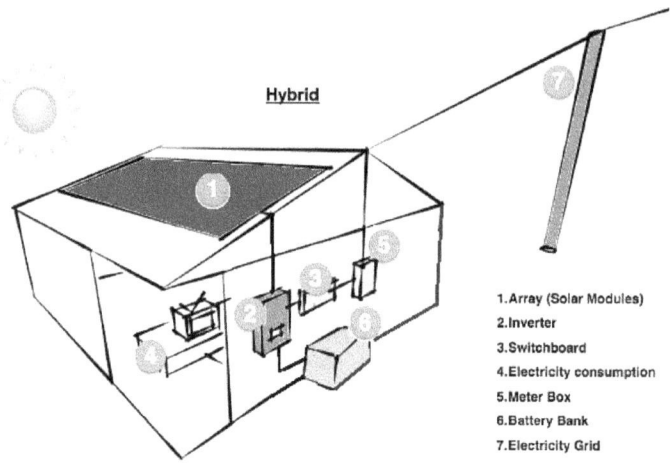

Hybrid

1. Array (Solar Modules)
2. Inverter
3. Switchboard
4. Electricity consumption
5. Meter Box
6. Battery Bank
7. Electricity Grid

128

Modern hybrid systems combine solar and battery storage in one and are now available in many different forms and configurations. Due to the decreasing cost of battery storage, systems that are already connected to the electricity grid can start taking advantage of battery storage as well. This means being able to store solar energy that is generated during the day and using it at night. When the stored energy is depleted, the grid is there as a back up, allowing consumers to have the best of both worlds. Hybrid systems are also able to charge the batteries using cheap off-peak electricity (usually after midnight to 6am).

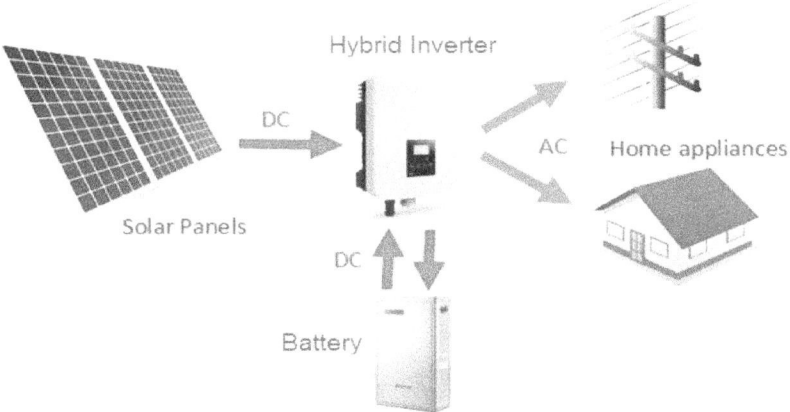

www.ingramcontent.com/pod-product-compliance
Lightning Source LLC
Chambersburg PA
CBHW072144170526
45158CB00004BA/1499

* 9 7 8 1 0 9 1 2 2 5 1 5 2 *